INTRO TO SPELEOLOGY & PALEONTOLOGY

Parent Lesson Planner
(PLP)

 Weekly Lesson Schedule

 Student Worksheets

 Quizzes & Test

Answer Key

7th – 9th grade

| 1 Year Science | 1/2 Credit |

First printing: March 2013
Second printing: August 2013

Master Books® is a division of the New Leaf Publishing Group, Inc.

ISBN: 978-0-89051-729-1

Unless otherwise noted, Scripture quotations are from the New King James Version of the Bible.

Printed in the United States of America

Please visit our website for other great titles:
www.masterbooks.net

For information regarding author interviews,
please contact the publicity department at (870) 438-5288

Master
Books®
A Division of New Leaf Publishing Group
www.masterbooks.net

Where Creation Inspires Education

Since 1975, Master Books has been providing educational resources based on a biblical worldview to students of all ages. At the heart of these resources is our firm belief in a literal six-day creation, a young earth, the global Flood as revealed in Genesis 1–11, and other vital evidence to help build a critical foundation of scriptural authority for everyone. By equipping students with biblical truths and their key connection to the world of science and history, it is our hope they will be able to defend their faith in a skeptical, fallen world.

If the foundations are destroyed, what can the righteous do?
Psalm 11:3; NKJV

As the largest publisher of creation science materials in the world, Master Books is honored to partner with our authors and educators, including:

Ken Ham of Answers in Genesis

Dr. John Morris and Dr. Jason Lisle of the Institute for Creation Research

Dr. Donald DeYoung and Michael Oard of the Creation Research Society

Dr. James Stobaugh, John Hudson Tiner, Rick and Marilyn Boyer, Dr. Tom Derosa, and so many more!

Whether a pre-school learner or a scholar seeking an advanced degree, we offer a wonderful selection of award-winning resources for all ages and educational levels.

But sanctify the Lord God in your hearts, and always be ready
to give a defense to everyone who asks you a reason for the hope
that is in you, with meekness and fear.
1 Peter 3:15; NKJV

Permission to Copy

Lessons for a 36-week course!

Overview: This *Introduction to Speleology and Paleontology PLP* contains materials for use with *The Cave Book* and *The Fossil Book* in the Wonders of Creation series. Materials are organized by each book in the following sections:

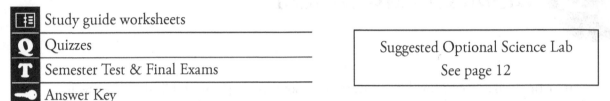

Study guide worksheets	
Q Quizzes	
T Semester Test & Final Exams	
Answer Key	

> Suggested Optional Science Lab
> See page 12

Features: Each suggested weekly schedule has two to three easy-to-manage lessons which combine reading, worksheets, and vocabulary-building opportunities including an expanded glossary for each book. Designed to allow your student to be independent, materials in this resource are divided by section so you can remove quizzes, tests, and answer keys before beginning the coursework. As always, you are encouraged to adjust the schedule and materials as you need to in order to best work within your educational program.

Workflow: Students will read the pages in their book and then complete each section of the course materials. Tests are given at regular intervals with space to record each grade. Younger students may be given the option of taking open book tests.

Lesson Scheduling: Space is given for assignment dates. There is flexibility in scheduling. For example, the parent may opt for a M-W schedule rather than a M, W, F schedule. Each week listed has five days but due to vacations the school work week may not be M-F. Please adapt the days to your school schedule. As the student completes each assignment, he/she should put an "X" in the box.

Approximately 30 to 45 minutes per lesson, two to three days a week	
Includes answer keys for worksheets, quizzes, and semester exams	
Worksheets for each chapter.	
Quizzes are included to help reinforce learning and provide assessment opportunities; optional semester exams included	
Designed for grades 7 to 9 in a one-year course to earn 1/2 science credit	
Suggested labs (if applicable)	

Course includes books from creationist authors with solid, biblical worldviews:

Emil Silvestru - *The Cave Book*

Dr. Emil Silvestru is a prospecting and exploration geologist. He is a cave expert born in Transylvania, Romania where he began exploring caves at the age of 12. He is currently a writer, researcher, and speaker.

Dr. Gary Parker - *The Fossil Book*

After starting his 30-year college teaching career as a non-Christian evolutionist, Dr. Gary Parker became a zealous creationist, eventually serving as professor of biology at the Institute for Creation Research in San Diego, lecturing worldwide for both ICR and Answers in Genesis, writing five science textbooks and seven creation books (translated into over 10 languages), and appearing in numerous films, videos and television programs.

Speleology Worksheets

for Use with

The Cave Book

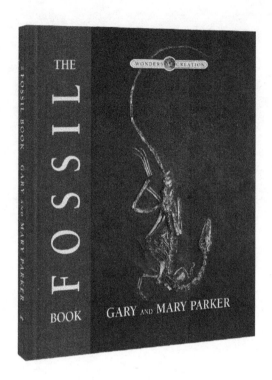

Paleontology Worksheets

for Use with

The Fossil Book

Quizzes & Tests Section

Define: (6 Points Each Answer)

1. Evolutionary series: _____

2. Metamorphosis: _____

3. Splint bones: _____

4. Vertebrates: _____

Multiple Answer Questions: (3 Points Each Blank)

5. What are the five groups of vertebrates?

   a. _____

   b. _____

   c. _____

   d. _____

   e. _____

6. Explain two biblical explanations for the extinction of dinosaurs.

   a. _____

   _____

   b. _____

   _____

Short Answer Questions: (4 Points Each Question)

7. Which of the following has the most DNA per cell of any other animal group?

   a. amphibians

   b. birds

   c. fish

   d. reptiles

   e. mammals

8. Coelacanths were once thought to be a transitional form between fish and amphibians. How was this claim disproved?

9. Explain what the Bible teaches about animals that eat meat today.

10. Discuss how dinosaurs can be explained by the biblical account of creation and the Flood.

11. What were the dimensions of the Ark?

12. Which dinosaurs were likely on Noah's Ark during the Flood?

13. Why should Archaeopteryx not be considered a missing link?

14. Explain why the alleged sequence of horse hooves does not prove evolution.

Applied Learning Activity: (6 Points Each Blank)

15. Discuss how the fossil record supports the biblical account of creation.

16. Discuss how the fossil record supports the biblical account of the corruption of the earth.

17. Discuss how the fossil record supports the biblical account of a global Flood.

Short Answer: (5 Points)

18. Discuss how the fossil record supports the biblical account of God's mercy on His creation.

Define: (2 Points Each Answer)

1. Evolution: _____

2. Paleontologist: _____

3. Permineralized fossils:_____

4. Living fossils: _____

5. Trilobite: _____

6. Cambrian explosion: _____

7. Arthropod: _____

8. Cephalopods: _____

9. Diatoms: _____

10. Evolutionary series: _____

11. Metamorphosis; _____

12. Splint bones; _____

Multiple Answer Questions: (2 Point Each Blank)

13. Who were the two men given credit for popularizing the modern teaching of evolution?

    a. _____

    b. _____

14. Who said that fossils are "perhaps the most obvious and serious objection to the theory of evolution"? Why is this significant?

    a. _____

    b. _____
    _____

15. What are the first animals fossilized in abundance? How does the complexity of these first fossils disprove evolution?

    a. _____

    b. _____
    _____

16. Discuss two biblical explanations for the extinction of dinosaurs.

    a. _____
    _____

    b. _____
    _____

17. What type of event would provide the right conditions to form fossils?

18. What catastrophic event in May of 1980 supports Dr. Austin's theory of how coal is formed?

19. According to Flood geologists, what does the geologic column show?

20. Explain why fossils of sea creatures are found throughout the geologic column while animals and land plants tend to be found higher in the column.

21. How do fossil clams testify to a rapid burial?

22. How did fossils of sea creatures end up on the top of mountains?

23. Why should Archaeopteryx not be considered a missing link?

24. Explain why the alleged sequence of horse hooves does not prove evolution.

Applied Learning Activity: (2 Points Each Answer)

List and describe the four C's of biblical history discussed in this book.

25. a.   b.

26. a.   b.

27. a.   b.

28. a.   b.

29. Fossils are found in geologic systems (such as the Cambrian), somewhat as living things are found in ecological zones (such as the ponds and woodlands of the hardwood forest zone). Perhaps geologic systems or paleosystems are the remains of pre-Flood ecological zones. Identify the twelve Geologic Column systems represented using: Cambrian, Cretaceous, Devonian, Jurassic, Mississippian, Ordovician, Pennsylvanian, Permian, Quarternary, Silurian, Tertiary, Triassic

Answer Keys

Introduction – Worksheet 1

karst — the term used by scientists to describe a landscape of caverns, sinking streams, sinkholes, and a vast array of small-scale features all generated by the solution of the bedrock, formed predominantly by limestones

karst aquifers — the assembly of ground water accumulated inside a karstic rock, enough to supply wells and springs

1. Their immediate need to find shelter from the rapidly cooling climate

2. Tower of Babel

3. 25 percent

4. Over 50 percent

1. It was deep inside the caves that some found shelter, mystical ritual hunting grounds, and a burial place for their dead.

2. The once-global knowledge and craftsmanship was split between many groups that could no longer truly communicate. Very quickly, various groups found themselves with the monopoly over one or several crafts/technologies, while other crafts were more or less lost for them. They were soon isolated from the other groups and many lost much of their knowledge of God.

Chapter 1 – Humans and Caves – Worksheet 1

acoustics — points of resonance (locations where if certain musical notes are emitted, they will bounce back, amplified, from the walls)

Acheulean industry — from the town of Saint-Acheul, whose most characteristic tool was the stone hand axe

bas-reliefs — artwork usually made of soft, pliable clay attached to walls or even to large blocks

cave paintings — either simple outlines of charcoal or mineral pigment, or true paintings with outlines, shading, and vivid pigment fills

engravings — usually made on soft limestone surfaces

Kyr — abbreviation for thousand years

Myr — abbreviation for million years

speleothems — mineral deposits that form inside caves; especially stalagmites and stalactites

1. Though we do not know for sure because there is no mention in Scripture, it is possible that there were caves prior to the Flood. They would have been formed differently than caves that exist today.

2. It is first mentioned in Genesis 19:30 concerning Lot and his daughters.

3. The word "cave" appears some 40 times in the Bible.

4. Cave bears, cave lions, and cave hyenas

5. China

6. Art associated with burial rituals

7. Paintings, engravings, and bas-reliefs

8. No, they were descended from the family of Noah.

1. Discussion might include their role as shelters or religious sanctuaries.

2. These early people carried their deep beliefs from their ancestor Noah. They also took on new beliefs as they separated from each other. Some may have come to see caves as an entrance into the earth. These were places of deep mystery to them.

3. The largest number of cave paintings are located in places of resonance (locations where if certain musical notes are emitted, they will bounce back, amplified, from the walls). It seems probable that chanting, dancing, and other types of ritual musical activities were associated with cave paintings.

4. Civilization; individual and unique. Man was created by God in His own image and was very intelligent and skilled from the beginning.

5. First representative of this human type was discovered in 1856 in a cave in the Neander Valley in Germany. Some have seen the remains as those belonging to an idiot, a hermit, or a medieval Mongolian warrior. Evolutionists were looking for a missing link, seeing this as a possible connection. However, they were simply humans with stocky, shorter bodies than many people today. They had broad noses and their brain size was slightly larger than that of modern humans.

6. Neanderthals had a spoken language, seemed to care for each other (those injured), and used flowers to decorate those buried.

Chapter 2 – Caves and Mythology – Worksheet 1

anthropods — invertebrate animals having an exoskeleton, segmented body and jointed appendages

bidirectional air circulation — air flowing two ways

cul-de-sac — cave with only one entrance

echolocation — bats send out sound waves that hit an object and an echo comes back, helping them identify the object

troglobites — creatures which live only in caves (from Greek for "cave dwellers")

troglophiles — creatures which spend some part of their life in caves (from Greek for "who like caves")

trogloxenes — creatures that got into a cave by accident and which try to leave (from Greek for "foreign to caves")

unidirectional air circulation — air flowing one way

1. Egypt, Phoenicia, Assyro-Babylonia, Greece, Rome, and Maya

2. The cave olm

3. Troglophiles

4. A spectacular cave environment where several new species of creatures were found

5. Usually about 90 percent

1. Often one or two other females spread their wings underneath the delivering mother, ready to catch the little one if needed.

2. Thirty-eight Ukrainian Jews hid during World War II for nearly two years.

3. Ice can accumulate in cul-de-sac shafts because they act as traps for cold air.

4. Some caves have an abundance of negative ions in the air, which are usually oxygen atoms. Someone with a cold or flu can improve more quickly because of the absence of cosmic radiation.

Chapter 3 – Caves and Karst – Worksheet 1

cave — considered a natural opening in rocks, accessible to humans, which is longer than it is deep and is at least 33 feet in length

emergences — the place where subterranean waters emerge to the surface

endogenetic — internal processes that can create caves

exogenetic — external process that can create caves

karsted — rich in karst features, especially caves

orthokarst — karst formed on carbonate rocks mainly by solution

parakarst — karst-like features formed on non-carbonate rocks, mainly by solution

pseudokarst — karst-like features formed on any kind of rock by other ways than solution

resurgences — the place where a sinking stream re-emerges to the surface

sinkholes — funnel-shaped hollows, from a few feet to hundreds of feet in diameter

1. Soluble rocks on which most landforms are formed by solution (karren, sinkholes, blind valleys, swallets, uvalas, poljes, etc.)

2. 12 percent

3. Evaporite rocks (rock gypsum and rock salt) and chalk

4. The exit point(s) of cave waters of a known stream, also called karst springs

5. Karst springs are called emergences when there is no evidence of the origin of the waters that emerge.

6. Up to 4,060.7 cubic feet per second; 115 tons of water every second, enough to supply the needs of more than two New Yorks every day.

7. Rhythmic springs flow intermittently, due to the very special shape and spatial distribution of the caves and conduits involved, as well as the constancy of water supply to the caves; Fontaine de Fontestorbes in southern France

1. Endogenetic caves are formed within moving lava. Lava tubes form when and where there are long-term lava flows. Often, stalactites also form. Exogenetic caves are the result of either chemical processes or physical processes, as volcanic ash and other pyroclastics are deposited.

2. Named by Austrian geographers and geologists in the 19th century while studying limestone terrains; they Germanized the Slovenian name "Kras" used by the locals. Probably comes from an old pre-Indo-European root "kara" meaning "desert of stone."

3. It might be riddled with all sorts of runnels, grooves, and small hollows called karren. Funnel-shaped hollows called sinkholes fill the terrain. Also, large hollows (depressions) called polje fill the karst terrain.

4. Northeastern South America (Venezuela and British Guiana): quartzite sandstone.

5. Many scientists agree today that this area was the result of the limestone being dissolved by sulfuric acid rising (not seeping down as in the case of proper karsting).

Chapter 4 – Classifying Caves – Worksheet 1

active caves — live caves that have a flowing stream in them

compoundrelict caves — fossil caves above the water table

detrital formations — sediments brought into the caves by streams and residual material left by the limestone

dripping speleothems — stalactites, stalagmites, and columns that are growing

phreatic caves — flooded (water saturated) caves formed and/or located below the water table

relict caves — caves without a flowing stream, which may have ponds or dripping water

denudation rate — the pace at which a given surface of bare rock is eroded; usually measured in millimeters per millennium (thousand years)

vadose caves — caves that formed and continue to exist mostly above the water table; the majority of their passages have air above the water

1. A live cave which has a flowing stream

2. Three: inflow, outflow, and through caves

3. A special type of speleothem made up of two semicircular plates growing parallel to each other. They can grow more than three feet in diameter.

4. Two kinds; those that grow from stalactites (helictites) and those that grow from stalagmites (heligmites)

5. When cave pool water is saturated and very calm, thin flakes of calcite start growing, floating on the water, and are called cave rafts.

6. Any three: gypsum flowers, balloons, crusts, chandeliers, angel hair (mirabilite), moonmilk

1. Speleothems are nearly pure calcite that was removed from the limestone and redeposited inside caves. When removed from the limestone, some other soluble minerals accompany calcite, and some contain the radioisotope uranium 234. This decays through a long series of intermediates into lead 206. This is calculated based on a certain rate of decay.

2. Recorded growths of speleothems refute the necessity of tens or hundreds of thousands of years.

Chapter 5 – Exploring Caves – Worksheet 1

desiccation cracks — cracks occurring because of shrinking of sediment as it dries

master joints — a tectonic discontinuity (fault line) that a given cave passage follows

scallops — spoon-shaped hollows in a cave wall, floor, or ceiling dissolved by eddies in flowing water

1. Alexander the Great in 325 B.C.

2. Jaques-Yves Cousteau

3. The Stermers (a Jewish family from Ukraine)

4. Popowa Yama Cave

1. Good footwear (hiking boots or rubber boots), good rope, flashlights, hard hat, gloves, pack, wool socks, and knee pads

2. This would include ascending and descending by rope, crawling through muck and water, moving through large and small chambers, following master joints and scallops.

3. Humidity increases the absorption of light; unpleasant colors can result when using artificial light films; much time is required to take simple shots with artificial lights.

Chapter 6 – Studying Caves – Worksheet 1

cenote — steep-walled natural well reaching the water table and continuing below it

concavities — corresponding niches to convexities

convexities — a vertical succession of ledges

diagenesis — a complex set of transformations through which sediments go, from compaction, through dewatering to cementation

xibalba — the Mayan "Underworld"

1. The karst denudation rate
2. The precipitation water and the fluvial water
3. Hydrograph; chemograph
4. As a transport and storage device
5. About 500 years

1. Subterranean geomorphology deals with the complex morphologies encountered under the ground and their relationship with the surface; geology studies the survey of all formations encountered to tectonics; geochemistry studies the direct chemical interactions of rocks with the environment; hydrology helps in understanding how waters move or are stored in the rocks; hydrogeology combines hydrology, geology, and chemistry.

2. Water has a way of finding its way through karst aquifers, draining away from the artificial water reservoir.

3. What happens on the surface has a significant effect on what happens under the ground because infiltration from surface water can be extremely fast.

Introduction – Solving the Fossil Mystery – Worksheet 1

Evolution — the belief that life started by chance, and millions of years of struggle and death slowly changed a few simple living things into many varied and complex forms through stages

Paleontologist — a person who studies fossils

1. During the 1600s and 1700s

2. True

3. Charles Lyell and Charles Darwin

4. Time, chance, struggle, and death

5. Creation (God created all things in six actual days about 6,000 years ago. The completed creation was "very good"), corruption (Adam's sin ushered death, disease, sickness, pain, etc. into the world), catastrophe (God judged the wickedness of mankind with a global, earth-covering flood during Noah's day, around 4,500 years ago), and Christ (Jesus Christ came to earth to redeem mankind from the curse of sin and death).

Chapter 1 – Fossils, Flooding, and Sedimentary Rock – Worksheet 1

Archaeology — the science that deals with human artifacts, and with things deliberately buried by humans

Artifacts — products crafted by humans

Fossil — remains or trace of a once-living thing preserved by natural processes, most often by rapid, deep burial in waterlaid sediments.

Geology — the scientific study of the earth, including the materials that it is made of, the physical and chemical processes that occur on its surface and in its interior, and the history of the planet and its life forms

Paleontology — the study of fossils

Permineralized fossils — fossils preserved by minerals hardening in the pore spaces of a specimen such as a shell, bone, or wood

Polystrates — fossils that cut through many layers, suggesting the sequence was laid down very rapidly

Pseudofossils — false fossils; things that look like fossils but really aren't

Sediments — particles of sand, silt, clay, ash, etc. eroded and deposited by wind and water currents

Trace fossils — are not remains of plant or animal parts, but show evidence of once-living things

1. Flaky shale, gritty sandstone, or chalky limestone

2. Wind and water

3. Water

4. Water and rock cement

5. False. Rocks and fossils can form quickly given the right conditions. Long periods of time are not needed to form rocks and fossils.

6. Calcium carbonate and silica

7. Any three: limestone, bottom of tea kettle, in Tums and Rolaids, chalk

8. Silica gel packs are placed in boxes of electronic equipment.

9. A flood

10. When a plant is buried in sediment under flood conditions, the plant is preserved when the heavy sediment weight squeezes out extra water and encourages the growth of cement minerals that turn the plant into a fossil.

11. The plant or animal needs to be preserved quickly before it begins to decompose.

12. A permineralized fossil

13. Permineralized wood has minerals in its pore spaces but still has wood fibers, while minerals have completely replaced the wood but preserved the pattern in petrified wood.

14. Coal is the charred remains and carbon atoms of once-living plants, making it a fossil. Coal burns, making it a fuel.

15. Huge mats of vegetation were ripped up in violent storms, torn apart by the waves and currents, and deposited in layers. Sediment on top of these layers then squeezed out water and raised the temperature of the buried plants. The plants would then begin to char, turning into coal.

16. The eruption of Mount St. Helens

17. If the layers surrounding the polystrate item had built up slowly over millions of years, the tops of the polystrate item would rot away, even if the bottoms were fossilized.

Chapter 2 – Geologic Column Diagram – Worksheet 1

Index fossil — fossils used to identify a geologic system because they lived either (a) at a certain time or (b) in a certain place in the pre-Flood world

Geologic column — a columnar diagram identifying rocky layers (strata) that form a sequence from bottom to top to indicate their relation to the twelve geologic systems

Living fossils — creatures found alive today that evolutionists thought became extinct millions of years ago

Trilobite — a crab-like creature that was the first fossil found buried in abundance around the world

1. Sedimentary rocks (limestone, shale, sandstone); cliffs, cuts, creeks, and quarries

2. Stages of evolutionary development over millions of years

3. A series of burials

4. 12; 3

5. Since they were buried later in Noah's Flood, paleosystems with land plants and animals occur higher in the geologic column diagram than those with only sea creatures, but fossils of sea life occur in all geologic systems or eco-sedimentary zones since the Flood waters eventually covered all the land.

6. Flood geologists use the word "first" to refer to the first to be buried by the Flood. They use the term "last" to refer to the last to be buried in the Flood. Evolutionists use the word "first" to refer to the first to evolve, meaning that nothing lived before it did. They use the word "last" to refer to the last surviving of its kind before it evolved into something else or became extinct.

7. Charles Darwin; Charles Darwin realized that evolution needed viable evidence of transitions from one animal into another; without them, evolution could not be validated.

Chapter 3 – Flood Geology vs. Evolution – Worksheet 1

Cambrian explosion — the sudden appearance of a wide variety of complex life forms in the lowest rock layer with abundant fossils (Cambrian); considered a challenge to evolution, these may be the first organisms in a corrupted creation to be buried in Noah's flood

Cavitations — bubbles formed by surging waters

Paraconformities — a gap without erosion in the geologic column diagram; breaks the time sequence assumed by evolution, and may suggest fossils from different environments were rapidly buried by a lot of water, not a lot of time

Stromatolites — banded rock deposits formed by blue-greens growing in mossy mats on rocks in the tidal zone along the shore; the mats trap and then cement sand grains to form a mineral layer, continually building new layers on top of earlier ones

1. Cambrian

2. They tried to look for simple life forms in pre-Cambrian rock.

3. False. Jellyfish and segmented worms are anything but simple.

4. Two-thirds

5. The waters that burst out of the deep during Noah's Flood

6. If the millions-of-years scenario were true and erosion occurred gradually, the softer rock would be gone and the hard rock would stick up into the sediment above. However, the tilted layers have been sheared off in a nearly straight line.

7. The bottom layers were most likely formed in the years before the Flood and sheared off during the beginning stages of the Flood. The upper layers were set into place during the Flood. Tectonic activity pushed these layers up as the water receded from the earth's surface during the later stages of the Flood. Water became trapped by earthen dams, which finally broke years after the Flood and released water to tear away the earth's surface. These cascades of water followed the easiest path downhill, which is now where we see the Colorado River through the gorge of Grand Canyon.

8. Mount St. Helens

Chapter 4 – Kinds of Fossils I: Invertebrates – Worksheet 1

Arthropod — all creatures with jointed legs and a tough outside skeleton (exoskeleton) made of chitin: insects, crabs and shrimp, spiders, centipedes, and millipedes

Cephalopod — means "head-footed," since their tentacles come out of their heads, the most complex of all the invertebrates are the squid and octopus in the mollusk class

Diatom — microscopic, one-celled plants whose walls are decorated with glass ($SiO2$) in exquisite patterns; mined and sold as diatomaceous earth, which is used in filtering and abrasion

Echinoderms — meaning "spiny-skinned," members of the starfish/sea star group usually have bony plates and spines embedded

Gastropods — means "stomach-footed," since they walk on their stomachs; mollusk class which snails belong

Invertebrate — animals without backbones

Malacology — a branch of science is devoted to the study of mollusk shells

Mollusks — a large phylum of animals with thick, muscular bodies and a complex system of organs

Nautiloids — fossils with tapered, chambered shells; some are coiled like the modern nautilus, others are curved like bananas, and still others are straight, like ice cream cones

Palynology — the branch of paleontology that studies microscopic spores and pollen of plants

Protozoan — one-celled animal

Spicules — ponges that have hard skeletal structures of crystal-like spines

1. Seashells

2. True

3. Many fossil clams are found with both sides of the shell still together. That means the clam must have been buried so deeply and so fast that it couldn't even open its shell to burrow out.

4. Shelled squids

5. The sutures of nautiloids vary stratigraphically from smooth to wiggly to very wiggly. Evolutionists use this to "prove" simple to more complex lifeforms.

6. First, the series starts with a very complex animal at the bottom of the GCD. Second, the first-buried form is also the fittest, since it's the only survivor. Third, the animal never evolved from anything or into anything. Fourth, there are few suggestions and no agreement on the survival value of having a wiggly suture. Fifth, there are reversals of the sequence evolutionists expected.

7. Great Barrier Reef in Australia

8. False. Scientists have measured coral growth and found that the largest reef in the world could have formed in less than 4,000 years.

9. Mississippian layer

10. They show that those areas were at one time completely covered by water.

11. It is possible that the volcanic activity that accompanied the Flood released toxins into the water that prevented the decomposition of the insects. Silt and clay buried the insects and settled quickly in briefly quiet water, solidifying fast enough to prevent later currents from tearing apart the fragile specimens.

12. Trilobites

13. Evolution assumes that the earliest fossils, which are found in the lowest layers, would be the most primitive and least complex since they hadn't yet evolved into more complex beings. However, since these fossils reveal complex creatures of design, they disprove the idea that non-complex beings changed into complex beings.

14. The turbulent Flood waters covered the entire earth, including the high hills (Genesis 7:19). Then the mountains rose, and the valleys sank down (Psalm 104:8). At the end of the Flood, God raised up the layers that were below the sea, lifting sea-creature fossils even to the tops of earth's highest mountains.

Chapter 5 – Kinds of Fossils II: Vertebrates – Worksheet 1

Evolutionary series — a sequence of fossils that suggests how one kind of creature might have changed into another

Metamorphosis — the process of transformation from an immature form to an adult form in two or more distinct stages

Splint bones — modern one-toed horses actually keep parts of the two flanking toes as important leg support structures (not useless evolutionary leftovers)

Vertebrates — animals with backbones

1. Fish, amphibians, reptiles, birds, and mammals

2. A.

3. Live coelacanths were found in the Indian Ocean and near Indonesia. There were no elbow or wrist joints as evolutionists once claimed; the stiff fin was used for steering and swimming, not walking. Their organs worked more like those in a shark, not those in a frog. The fish did not live in ponds; it lived in the deep ocean.

4. God created all animals and people in the beginning to eat only plants. It was only after man's sin corrupted God's perfect creation that some animals began to eat other animals, and it was only after the

Flood that God gave mankind permission to eat meat.

5. Dinosaurs were created on Day 6 along with the other land animals and people. Two representatives of the various dinosaur kinds were on the Ark. Since the average height of dinosaurs was about the size of a small pony, and since younger dinosaurs were smaller than older ones, they would have fit on Noah's Ark during the global Flood. Those that weren't on the Ark perished in the Flood. Many were buried in the muddy sediments. Those that survived the Flood on the Ark repopulated the earth after disembarking, although most eventually died from various causes.

6. 450 feet long; 75 feet wide; 45 feet high (150 x 25 x 15 meters)

7. They were likely young adults since God desired them to replenish the earth after the Flood.

8. The first explanation is because climate and soil conditions changed; dinosaurs had a difficult time surviving in that "new" world. The second explanation is that they were over-hunted by people after the Flood. Both are certainly possibilities, but we are not absolutely certain why dinosaurs went extinct.

9. Features in Archaeopteryx that evolutionists claimed to be transitional were and are found in other birds. The conclusion was made that Archaeopteryx was just a strong flying bird. Additionally, regular birds are found lower in the geologic column than Archaeopteryx.

10. First, the claimed earliest horse fossil is actually still seen today as a hyrax, or coney, not a primitive horse. Second, the change in horse size is not support for evolution since extreme sizes can be achieved simply by breeding. Third, the difference in hoof number can be explained by variation in horse kinds that were better able to live in different environments. Finally, the three-hoofed and one-hoofed horse kinds lived at the same time. These animals are not evidence for evolutionary transition.

11. Answers will vary.

Conclusion – Worksheet 1

1. The first buried fossils of each group are complete and complex, with all the features that separate its kind from all the others.

2. Since Adam sinned, the earth and all that was in it was cursed. Fossils themselves are dead things. Death was not part of God's original creation; it came as a result of sin. The fossil record also testifies to animals eating other animals; this did not occur before sin since there was no death before sin and since God originally created man and animals to be vegetarian.

3. 1. Dead things are broken down so fast that most fossils must have formed rapidly or they wouldn't have formed at all.

   2. Most fossils are found in sedimentary rocks that form in the way concrete cures, so the right conditions form rock quickly and no amount of time can form rocks under the wrong conditions.

   3. Some dinosaur bones and other fossils contain DNA, protein, or other chemicals that would break down completely in thousands of years, not millions.

   4. Countless numbers of living things must have been buried at the same time and place to form oil deposits, and that must have happened no more than thousands of years ago, or the oil would have leaked to the surface.

   5. Gaps in the GCD with insufficient evidence of erosion, such as the "150 million missing years" in the walls of Grand Canyon, suggest evolution's millions of years are a myth.

   6. Misplaced fossils, like fossils of woody plants in Cambrian rock and living fossils, show that fossils from various geologic systems lived at the same time in different places, not at different times in the same place. The systems in the geologic column seem to be primarily the buried remains of different life zones in the pre-Flood world.

7. Scientists studying the 1980 and 1982 eruptions of Mount St. Helens saw powerful evidence that catastrophic processes can do in days what slow processes could never do, not even in millions of years.

4. God preserved His creation while enacting judgment on the world. Noah, his wife, his sons, and their wives, along with two of every kind of unclean land animal and seven of every kind of clean land animal, were preserved by God's grace on the Ark. He promised judgment and yet made a way to escape that judgment. After the Flood, God restored His creation to life again. Man reproduced after his kind, animals reproduced after their kind, and plants reproduced after their kind, just as God commanded and desired.

Fossil Application – Worksheet 1

1. Sedimentary rocks – limestones, sandstones, shale

2. Cliffs, cuts (for roads), creeks, quarries

3. No – some states and specific locations like national parks require special permits; others will allow you to look for fossils but you cannot collect them.

4. Always check with your state laws, but usually in the US, property owners are considered the owners of any fossils found on their property.

5. A lot of rock removal has already been done, and sometimes you can see fossils that have been exposed or easily find them in the rock remains from quarrying.

6. Quarries can be very unstable, and contain quicksand, rockfalls, deep water, sharp rocks, dangerous cliffs, and even dangerous creatures like snakes or alligators.

7. Answer will vary - can include box, bag, bucket; tissue, cloth, sealable or plastic bags

8. Wet screening is when a screen is used with a shovel or scoop that allows you to pour the sandy material onto the screen and then gently shake it in the water; this works best at a beach or even without water in sandy locations on land

9. Matrix

10. Answers will vary. Fossils are old and fragile — you could easily damage the fossil with the wrong tools or by getting too close to it.

11. A. Trilobites, B. Echinoderms, C. nautiliods, D. corals, E. Brachiopods, F. ferns, G. mollusks

12. By matching the size scale and other identifying visual features, you can begin to visually identify fossils in the field. You also may not recognize a fossil by scientific name but you can by sight.

13. No. Most fossils need only minimal cleaning; fragile or crumbly fossils can be stabilized by using a concrete sealer.

14. Match the fossil with the biblical application that can be made by drawing lines:
   a. Nautilus > Creation
   b. Trilobites or fossils with bite marks > Corruption
   c. Living Fossils > Christ
   d. Closed fossil clams > Catastrophe

Bonus Question: Answers will vary.

True/False Questions:

1. F, 2. F, 3. F, 4. T, 5. F, 6. F, 7. T, 8. F, 9. T, 10. F, 11. T, 12. F, 13. T, 14. T,

15. F (many show complex beginning like trilobites and nautiloids)

Introduction – Chapter 1 Quiz 1

1. **karst** — the term used by scientists to describe a landscape of caverns, sinking streams, sinkholes, and a vast array of small-scale features all generated by the solution of the bedrock, formed predominantly by limestones

2. **Acheulean industry** — from the town of Saint-Acheul, whose most characteristic tool was the stone hand axe

3. **bas-reliefs** — artwork usually made of soft, pliable clay attached to walls or even to large blocks

4. **Kyr** — abbreviation for thousand years

5. **Myr** — abbreviation for million years

6. **Neanderthals** — believed by some to be an early human, found in the Neander Valley ("Neader Thal" in German)

7. **speleothems** — mineral deposits that form inside caves; especially stalagmites and stalactites

8. **karst aquifers** — the assembly of ground water accumulated inside a karstic rock, enough to supply wells and springs

9. Any two: cave bears, cave lions, cave hyenas

10. Paintings, engravings, and bas-reliefs

11. Their immediate need to find shelter from the rapidly cooling climate

12. It was deep inside the caves that some found shelter, mystical ritual hunting grounds, and a burial place for their dead.

13. 25 percent

14. Over 50 percent

15. Tower of Babel

16. The once-global knowledge and craftsmanship was split between many groups that could no longer truly communicate. Very quickly, various groups found themselves with the monopoly over one or several crafts/technologies, while other crafts were more or less lost for them. They were soon isolated from the other groups and many lost much of their knowledge of God.

17. It is first mentioned in Genesis 19:30 concerning Lot and his daughters.

18. China

19. (c) art associated with burial rituals

20. No, they were descended from the family of Noah.

21. The largest number of cave paintings are located in places or resonance (locations where if certain musical notes or sounds are emitted, they will bounce back, amplified, from the walls). It seems probable that chanting, dancing, and other types of ritual musical activities were associated with cave paintings

Chapter 2-3 Quiz 2

1. **troglobites** — creatures which live only in caves (from Greek for "cave dwellers")

2. **troglophiles** — creatures which spend some part of their life in caves (from Greek for "who like caves")

3. **trogloxenes** — creatures that got into a cave by accident and which try to leave (from Greek for "foreign to caves")

4. **endogenetic** — internal processes that can create caves

5. **exogenetic** — external process that can create caves

6. **resurgences** — the place where a sinking stream re-emerges to the surface

7. Two years; they were hiding from the Nazis.

8. Any four: Egypt, Phoenicia, Assyro-Babylonia, Greece, Rome, and Maya

9. Rock gypsum, rock salt, chalk

10. The cave olm

11. (b) troglophiles

12. A spectacular cave environment where several new species of creatures were found

13. Soluble rocks on which most landforms are formed by solution (karren, sinkholes, blind valleys, swallets, uvalas, poljes, etc.)

14. 12 percent

15. Named by Austrian geographers and geologists in the 19th century while studying limestone terrains; they Germanized the Slovenian name "Kras" used by the locals. Probably comes from an old pre-Indo-European root "kara" meaning "desert of stone."

16. Rhythmic springs flow intermittently, due to the very special shape and spatial distribution of the caves and conduits involved, as well as the constancy of water supply to the caves.

17. Usually around 90 percent

18.

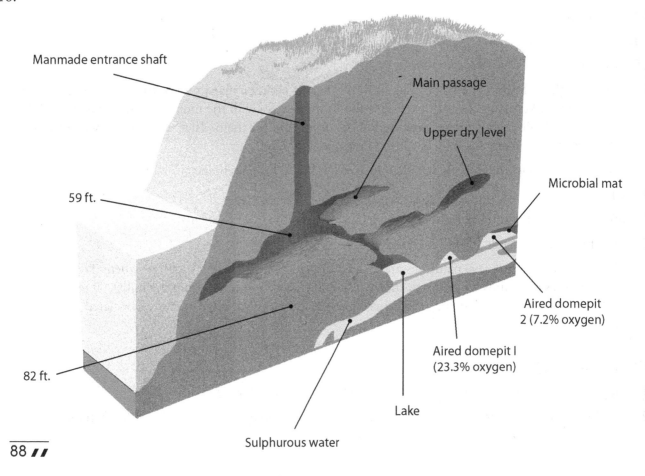

Manmade entrance shaft

Main passage

Upper dry level

Microbial mat

59 ft.

Aired domepit 2 (7.2% oxygen)

82 ft.

Aired domepit I (23.3% oxygen)

Lake

Sulphurous water

Chapter 4-5 Quiz 3

1. **detrital formations** — sediments brought into the caves by streams and residual material left by the limestone

2. **dripping speleothem** — stalactites, stalagmites, and columns that are growing

3. **desiccation cracks** — cracks occurring because of shrinking of sediment as it dries

4. **master joints** — a tectonic discontinuity (fault line) that a given cave passage follows

5. Inflow, outflow, and through caves

6. Those that grow from stalactites (helictites) and those that grow from stalagmites (heligmites)

7. Any three: gypsum flowers, balloons, crusts, chandeliers, angel hair (mirabilite), moonmilk

8. Any four: good footwear (hiking boots or rubber boots), good rope, flashlights, hard hat, gloves, pack, wool socks, and knee pads

9. Ascending and descending by rope, crawling through muck and water, moving through large and small chambers, following master joints and scallops.

10. a. The Stermers (a Jewish family from Ukraine) b. Popowa Yama Cave

11. a. Humidity increases the absorption of light;

    b. unpleasant colors can result when using artificial light films;

    c. much time is required to take simple shots with artificial lights.

12. A special type of speleothem made up of two semicircular plates growing parallel to each other. They can grow more than three feet in diameter.

13. When cave pool water is saturated and very calm, thin flakes of calcite start growing, floating on the water, and are called cave rafts.

14. Recorded growths of speleothems refute the necessity of tens or hundreds of thousands of years.

15. Alexander the Great in 325 B.C.

16. Jaques-Yves Cousteau

17.

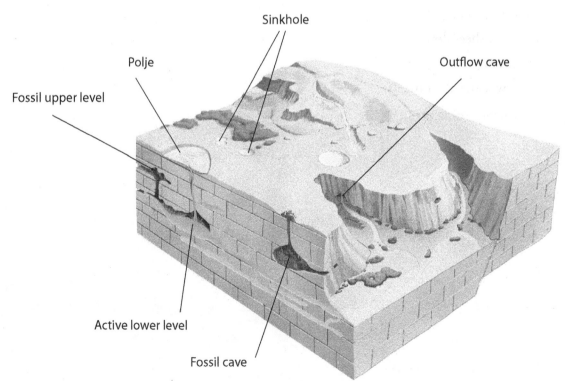

Chapter 6 Quiz 4

1. **cenote** — steep-walled natural well reaching the water table and continuing below it
2. **concavities** — corresponding niches to convexities
3. **convexities** — a vertical succession of ledges
4. **diagenesis** — a complex set of transformations through which sediments go, from compaction, through dewatering to cementation
5. **Xibalba** — the Mayan "Underworld"
6. The precipitation water and the fluvial water
7. Hydrograph; chemograph
8. Subterranean geomorphology, geology, geochemistry, hydrology, hydrogeology
9. The karst denudation rate
10. As a transport and storage device
11. About 500 years
12. Water has a way of finding its way through karst aquifers, draining away from the artificial water reservoir.
13. What happens on the surface has a significant effect on what happens under the ground because infiltration from surface water can be extremely fast.
14. calcite
15. Answers may vary but should include the thought that it doesn't take millions of years for calcite to cover an item, only the right conditions.
16. Stage 1: a: insoluble rocks
    b: soluble rocks
    c: hyperactive hydrothermal solutions generated during the Flood
    d: Large karst cavities excavated immediately after diagenesis

    Stage 2: e: global tectonic uplift
    f: global sheet flow
    g: massive rain
    h: new, detrital sediments

    Stage 3: e: karstic sediments

1. **bas-reliefs** — artwork usually made of soft, pliable clay attached to walls or even to large blocks

2. **Kyr** — abbreviation for thousand years

3. **Myr** — abbreviation for million years

4. **troglobites** — creatures which live only in caves (from Greek for "cave dwellers")

5. **troglophiles** — creatures which spend some part of their life in caves (from Greek for "who like caves")

6. **trogloxenes** — creatures that got into a cave by accident and which try to leave (from Greek for "foreign to caves")

7. **detrital formations** — sediments brought into the caves by streams and residual material left by the limestone

8. **dripping speleothem** — stalactites, stalagmites, and columns that are growing

9. **desiccation cracks** — cracks occurring because of shrinking of sediment as it dries

10. **concavities** — corresponding niches to convexities

11. **convexities** — a vertical succession of ledges

12. **diagenesis** — a complex set of transformations through which sediments go, from compaction, through dewatering to cementation

13. Paintings, engravings, and bas-reliefs

14. rock gypsum, rock salt, chalk

15. Any four: good footwear (hiking boots or rubber boots), good rope, flashlights, hard hat, gloves, pack, wool socks, and knee pads

16. Hydrograph; chemograph

17. Tower of Babel

18. No, they were descended from the family of Noah.

19. (b) troglophiles

20. A spectacular cave environment where several new species of creatures were found

21. Recorded growths of speleothems refute the necessity of tens or hundreds of thousands of years.

22. Alexander the Great in 325 B.C.

23. About 500 years

24. Water has a way of finding its way through karst aquifers, draining away from the artificial water reservoir.

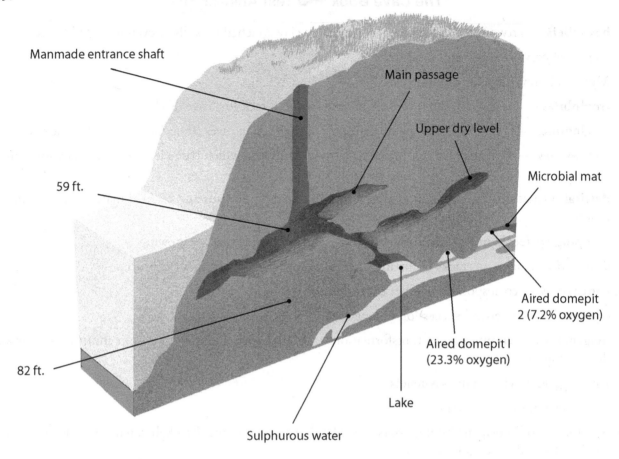

Manmade entrance shaft

Main passage

Upper dry level

Microbial mat

59 ft.

Aired domepit 2 (7.2% oxygen)

Aired domepit I (23.3% oxygen)

82 ft.

Lake

Sulphurous water

troduction – Chapter 1 Quiz 1

Evolution — the belief that life started by chance, and millions of years of struggle and death slowly changed a few simple living things into many complex and varied forms through stages

Paleontologist — a person who studies fossils

Permineralized fossils — fossils preserved by minerals hardening in the pore spaces of a specimen such as a shell, bone, or wood

Trace fossils — are not remains of plant or animale parts, but show evidence of once-living things

Charles Lyell and Charles Darwin

Wind and water; Water

Water and rock cement

Calcium carbonate and silica

Any three: limestone, bottom of tea kettle, in Tums and Rolaids, chalk

). During the 1600s and 1700s

1. Time, chance, struggle, and death

2. Flaky shale, gritty sandstone, or chalky limestone

3. A flood

4. When a plant is buried in sediment under flood conditions, the plant is preserved when the heavy sediment weight squeezes out extra water and encourages the growth of cement minerals that turn the plant into a fossil.

5. The plant or animal needs to be preserved quickly before it begins to decompose.

6. A permineralized fossil

7. Permineralized wood has minerals in its pore spaces but still has wood fibers, while minerals have completely replaced the wood but preserved the pattern in petrified wood.

8. Coal is the charred remains and carbon atoms of once-living plants, making it a fossil. Coal burns, making it a fuel.

). Huge mats of vegetation were ripped up in violent storms, torn apart by the waves and currents, and deposited in layers. Sediment on top of these layers then squeezed out water and raised the temperature of the buried plants. The plants would then begin to char, turning into coal.

). The eruption of Mount St. Helens

. If the layers surrounding the polystrate item had built up slowly over millions of years, the tops of the polystrate item would rot away, even if the bottoms were fossilized.

2. a. Creation b.God's perfect creation

. a. Corruption b. Ruined by man's sin

. a. Catastrophe b. Destroyed by Noah's flood

. a. Christ b. Restored to life in Christ

Chapter 2-3 Quiz 2

1. **Index fossils** — fossils used to identify a geologic system because they lived either (a) at a certain time o (b) in a certain place in the pre-Flood world

2. **Geologic column** — a columnar diagram identifying rocky layers (strata) that form a sequence from bottom to top to indicate their relation to the twelve geologic systems

3. **Living fossils** — creatures found alive today that evolutionists thought became extinct millions of years ago

4. **Trilobite** — a crab-like creature was the first fossil found buried in abundance around the world

5. **Cambrian explosion** — the sudden appearance of a wide variety of complex life forms in the lowest rock layer with abundant fossils (Cambrian); considered a challenge to evolution, these may be the first organisms in a corrupted creation to be buried in Noah's flood

6. **Cavitations** — bubbles formed by surging waters

7. **Paraconformities** — a gap without erosion in the geologic column diagram; breaks the time sequence assumed by evolution, and may suggest fossils from different environments were rapidly buried by a lot of water, not a lot of time.

8. **Stromatolites** — banded rock deposits formed by blue-greens growing in mossy mats on rocks in the tidal zone along the shore; the mats trap and then cement sand grains to form a mineral layer, continually building new layers on top of earlier ones

9. a. Sedimentary rocks (limestone, shale, sandstone); b. cliffs, cuts, creeks, and quarries

10. 12; 3

11. Charles Darwin; Charles Darwin realized that evolution needed viable evidence of transitions from one animal into another; without them, evolution could not be validated.

12. A series of burials

13. Since they were buried later in Noah's Flood, paleosystems with land plants and animals occur higher in the geologic column diagram than those with only sea creatures, but fossils of sea life occur in all geologic systems or eco-sedimentary zones since the Flood waters eventually covered all the land.

14. The waters that burst out of the deep during Noah's Flood

15. If the millions-of-years scenario were true and erosion occurred gradually, the softer rock would be gone and the hard rock would stick up into the sediment above. However, the tilted layers have been sheared off in a nearly straight line.

16. The bottom layers were most likely formed in the years before the Flood and sheared off during the beginning stages of the Flood. The upper layers were set into place during the Flood. Tectonic activity pushed these layers up as the water receded from the earth's surface during the later stages of the Flood. Water became trapped by earthen dams, which finally broke years after the Flood and released water to tear away the earth's surface. These cascades of water followed the easiest path downhill, which is now where we see the Colorado River through the gorge of Grand Canyon.

17. Mount St. Helens

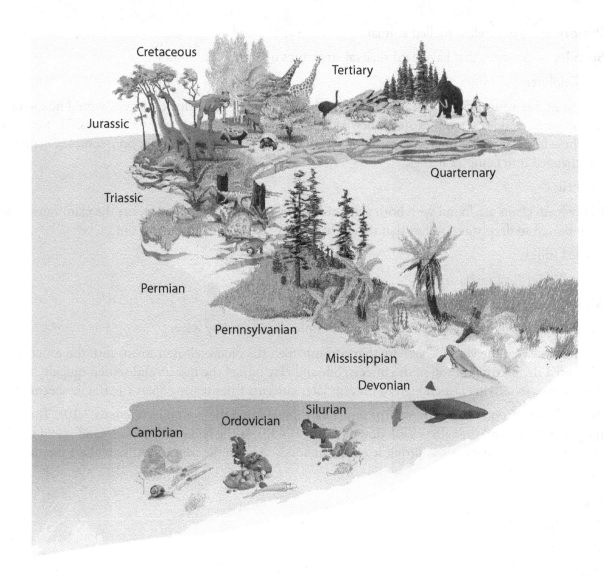

Cretaceous

Tertiary

Jurassic

Triassic

Quarternary

Permian

Pernnsylvanian

Mississippian

Devonian

Silurian

Cambrian

Ordovician

Chapter 4 Quiz 3

Arthropod — all creatures with jointed legs and a tough outside skeleton (exoskeleton) made of chitin: insects, crabs and shrimp, spiders, centipedes, and millipedes

Cephalopods — means "head-footed," since their tentacles come out of their heads; the most complex of all the invertebrates are the squid and octopus in the mollusk class

Diatoms — microscopic, one-celled plants whose walls are decorated with glass (SiO_2) in exquisite patterns; mined and sold as diatomaceous earth, which is used in filtering and abrasion

Echinoderms — meaning "spiny-skinned," members of the starfish/sea star group usually have bony plates and spines embedded

Gastropods — means "stomach-footed," since they walk on their stomachs; mollusk class which snails belong

Malacology — a branch of science is devoted to the study of mollusk shells

Nautiloids — fossils with tapered, chambered shells; some are coiled like the modern nautilus, others are curved like bananas, and still others are straight, like ice cream cones

Palynology — the branch of paleontology that studies microscopic spores and pollen of plants

9. **Protozoan** — one-celled shelled animals

10. **Spicules** — Sponges that have hard skeletal structures of crystal-like spines

11. a. Trilobites

   b. Evolution assumes that the earliest fossils, which are found in the lowest layers, would be the most primitive and least complex since they hadn't yet evolved into more complex beings. However, since these fossils reveal complex creatures of design, they disprove the idea that non-complex beings changed into complex beings.

12. Seashells

13. Many fossil clams are found with both sides of the shell still together. That means the clam must have been buried so deeply and so fast that it couldn't even open its shell to burrow out.

14. Shelled squids

15. Great Barrier Reef in Australia

16. Mississippian layer

17. They show that those areas were at one time completely covered by water.

18. It is possible that the volcanic activity that accompanied the Flood released toxins into the water that prevented the decomposition of the insects. Silt and clay buried the insects and settled quickly in briefly quiet water, solidifying fast enough to prevent later currents from tearing apart the fragile specimens.

19. The turbulent Flood waters covered the entire earth, including the high hills (Genesis 7:19). Then the mountains rose, and the valleys sank down (Psalm 104:8). At the end of the Flood, God raised up the layers that were below the sea, lifting sea-creature fossils even to the tops of earth's highest mountains.

20. Head

21. Eyes

22. Body (thorax)

23. Tail

24. Tail

25. Head with cheeks

26. Thorax

Chapter 5 - Conclusion Quiz 4

1. **Evolutionary series** — a sequence of fossils that suggests how one kind of creature might have changed into another

2. **Metamorphosis** — the process of transformation from an immature form to an adult form in two or more distinct stages

3. **Splint bones** — modern one-toed horses actually keep parts of the two flanking toes as important leg support structures (not useless evolutionary leftovers)

4. **Vertebrates** — animals with backbones

5. Fish, amphibians, reptiles, birds, and mammals

6. The first explanation is because climate and soil conditions changed, dinosaurs had a difficult time surviving in that "new" world.

   The second explanation is that they were over-hunted by people after the Flood. Both are certainly possibilities, but we are not absolutely certain why dinosaurs went extinct.

7. a. amphibians

8. Live coelacanths were found in the Indian Ocean and near Indonesia. There were no elbow or wrist joints as evolutionists once claimed; the stiff fin was used for steering and swimming, not walking. Their organs worked more like those in a shark, not those in a frog. The fish did not live in ponds; it lived in the deep ocean.

9. God created all animals and people in the beginning to eat only plants. It was only after man's sin corrupted God's perfect creation that some animals began to eat other animals, and it was only after the Flood that God gave mankind permission to eat meat.

10. Dinosaurs were created on Day 6 along with the other land animals and people. Two representatives of the various dinosaur kinds were on the Ark. Since the average height of dinosaurs was about the size of a small pony, and since younger dinosaurs were smaller than older ones, they would have fit on Noah's Ark during the global Flood. Those that weren't on the Ark perished in the Flood. Many were buried in the muddy sediments. Those that survived the Flood on the Ark repopulated the earth after disembarking, although most eventually died from various causes.

11. 450 feet long; 75 feet wide; 45 feet high (150 x 25 x 15 meters)

12. They were likely young adults since God desired them to replenish the earth after the Flood.

13. Features in Archaeopteryx that evolutionists claimed to be transitional were and are found in other birds. The conclusion was made that Archaeopteryx was just a strong flying bird. Additionally, regular birds are found lower in the geologic column than Archaeopteryx.

14. First, the claimed earliest horse fossil is actually still seen today as a hyrax, or coney, not a primitive horse. Second, the change in horse size is not support for evolution since extreme sizes can be achieved simply by breeding. Third, the difference in hoof number can be explained by variation in horse kinds that were better able to live in different environments. Finally, the three-hoofed and one-hoofed horse kinds lived at the same time. These animals are not evidence for evolutionary transition.

15. The first buried fossils of each group are complete and complex, with all the features that separate its kind from all the others.

16. Since Adam sinned, the earth and all that was in it was cursed. Fossils themselves are dead things. Death was not part of God's original creation; it came as a result of sin. The fossil record also testifies to animals eating other animals; this did not occur before sin since there was no death before sin and since God originally created man and animals to be vegetarian.

17. See the seven explanations on pages 69–70.

    1. Dead things are broken down so fast that most fossils must have formed rapidly or they wouldn't have formed at all.

    2. Most fossils are found in sedimentary rocks that form in the way concrete cures, so the right conditions form rock quickly and no amount of time can form rocks under the wrong conditions.

    3. Some dinosaur bones and other fossils contain DNA, protein, or other chemicals that would break down completely in thousands of years, not millions.

    4. Countless numbers of living things must have been buried at the same time and place to form oil deposits, and that must have happened no more than thousands of years ago, or the oil would have leaked to the surface.

    5. Gaps in the GCD with insufficient evidence of erosion, such as the "150 million missing years" in the walls of Grand Canyon, suggest evolution's millions of years are a myth.

6. Misplaced fossils, like fossils of woody plants in Cambrian rock and living fossils, show that fossils from various geologic systems lived at the same time in different places, not at different times in the same place. The systems in the geologic column seem to be primarily the buried remains of different life zones in the pre-Flood world.

7. Scientists studying the 1980 and 1982 eruptions of Mount St. Helens saw powerful evidence that catastrophic processes can do in days what slow processes could never do, not even in millions of years.

18. God preserved His creation while enacting judgment on the world. Noah, his wife, his sons, and their wives, along with two of every kind of unclean land animal and seven of every kind of clean land animal, were preserved by God's grace on the Ark. He promised judgment and yet made a way to escape that judgment. After the Flood, God restored His creation to life again. Man reproduced after his kind, animals reproduced after their kind, and plants reproduced after their kind, just as God commanded and desired.

The Fossil Book —🔑 Test Answer Key

1. **Evolution** — the belief that life started by chance, and millions of years of struggle and death slowly changed a few simple living things into many complex and varied forms through stages

2. **Paleontologist** — a person who studies fossils

3. **Permineralized fossils** — fossils preserved by minerals hardening in the pore spaces of a specimen such as a shell, bone, or wood

4. **Living fossils** — creatures found alive today that evolutionists thought became extinct millions of years ago

5. **Trilobite** — a crab-like creature was the first fossil found buried in abundance around the world

6. **Cambrian explosion** — the sudden appearance of a wide variety of complex life forms in the lowest rock layer with abundant fossils (Cambrian); considered a challenge to evolution, these may be the first organisms in a corrupted creation to be buried in Noah's flood

7. **Arthropod** — all creatures with jointed legs and a tough outside skeleton (exoskeleton) made of chitin: insects, crabs and shrimp, spiders, centipedes, and millipedes

8. **Cephalopods** — means "head-footed," since their tentacles come out of their heads; the most complex of all the invertebrates are the squid and octopus in the mollusk class

9. **Diatoms** — microscopic, one-celled plants whose walls are decorated with glass ($SiO2$) in exquisite patterns; mined and sold as diatomaceous earth, which is used in filtering and abrasion

10. **Evolutionary series** — a sequence of fossils that suggests how one kind of creature might have changed into another

11. **Metamorphosis** — the process of transformation from an immature form to an adult form in two or more distinct stages

12. **Splint bones** — Modern one-toed horses actually keep parts of the two flanking toes as important leg support structures (not useless evolutionary leftovers)

13. Charles Lyell and Charles Darwin

14. Charles Darwin; Charles Darwin realized that evolution needed viable evidence of transitions from one animal into another; without them, evolution could not be validated.

15. a. Trilobites

b. Evolution assumes that the earliest fossils, which are found in the lowest layers, would be the most primitive and least complex since they hadn't yet evolved into more complex beings. However, since these fossils reveal complex creatures of design, they disprove the idea that non-complex beings changed into complex beings.

16. The first explanation is because climate and soil conditions changed, dinosaurs had a difficult time surviving in that "new" world.
The second explanation is that they were over-hunted by people after the Flood. Both are certainly possibilities, but we are not absolutely certain why dinosaurs went extinct.

17. A flood

18. The eruption of Mount St. Helens

19. A series of burials

20. Since they were buried later in Noah's Flood, paleosystems with land plants and animals occur higher in the geologic column diagram than those with only sea creatures, but fossils of sea life occur in all geologic systems or eco-sedimentary zones since the Flood waters eventually covered all the land.

21. Many fossil clams are found with both sides of the shell still together. That means the clam must have been buried so deeply and so fast that it couldn't even open its shell to burrow out.

22. The turbulent Flood waters covered the entire earth, including the high hills (Genesis 7:19). Then the mountains rose, and the valleys sank down (Psalm 104:8). At the end of the Flood, God raised up the layers that were below the sea, lifting sea-creature fossils even to the tops of earth's highest mountains.

23. Features in Archaeopteryx that evolutionists claimed to be transitional were and are found in other birds. The conclusion was made that Archaeopteryx was just a strong flying bird. Additionally, regular birds are found lower in the geologic column than Archaeopteryx.

24. First, the claimed earliest horse fossil is actually still seen today as a hyrax, or coney, not a primitive horse. Second, the change in horse size is not support for evolution since extreme sizes can be achieved simply by breeding. Third, the difference in hoof number can be explained by variation in horse kinds that were better able to live in different environments. Finally, the three-hoofed and one-hoofed horse kinds lived at the same time. These animals are not evidence for evolutionary transition.

25. a. Creation b. God's perfect creation

26. a. Corruption b. Ruined by man's sin

27. a. Catastrophe b. Destroyed by Noah's flood

28. a. Christ b. Restored to life in Christ

29.

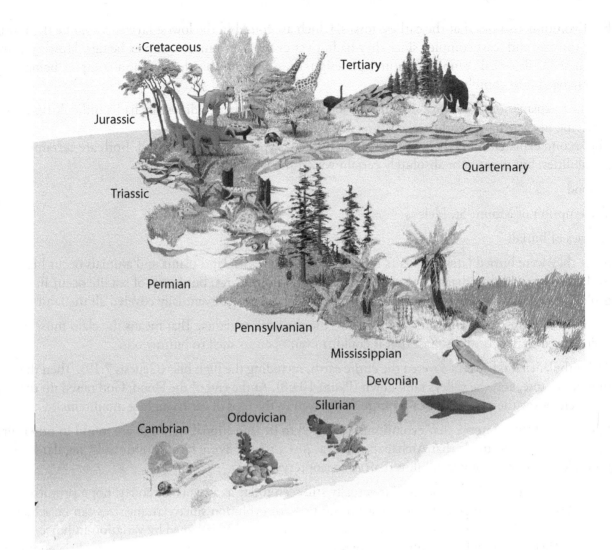

The Cave Book - Glossary

Acoustics — points of resonance (locations where if certain musical notes are emitted, they will bounce back, amplified, from the walls)

Acheulean industry — from the town of Saint-Acheul, whose most characteristic tool was the stone hand axe

Active caves — live caves that have a flowing stream in them

Arthropods — invertebrate animals having an exoskeleton, segmented body and jointed appendages

Bas-reliefs — artwork usually made of soft, pliable clay attached to walls or even to large blocks

Bidirectional air circulation — air flowing two ways

Cave — considered a natural opening in rocks, accessible to humans, which is longer than it is deep and is at least 33 feet in length

Cave paintings — either simple outlines of charcoal or mineral pigment, or true paintings with outlines, shading, and vivid pigment fills

Cenote — steep-walled natural well reaching the water table and continuing below it

Compoundrelict caves — fossil caves above the water table

Concavities — corresponding niches to convexities

Convexities — a vertical succession of ledges

Cul-de-sac — cave with only one entrance

Denudation rate — the pace at which a given surface of bare rock is eroded, usually measured in millimeters per millennium (thousand years)

Desiccation cracks — cracks occurring because of shrinking of sediment as it dries

Detrital formations — sediments brought into the caves by streams and residual material left by the limestone

Diagenesis — a complex set of transformations through which sediments go, from compaction, through dewatering to cementation

Dripping speleothems — stalactites, stalagmites, and columns that are growing

Echolocation — bats send out sound waves that hit an object and an echo comes back, helping them identify the object

Emergences — the place where subterranean waters emerge to the surface

Endogenetic — internal processes that can create caves

Engravings — usually made on soft limestone surfaces

Exogenetic — external process that can create caves

Karst — the term used by scientists to describe a landscape of caverns, sinking streams, sinkholes, and a vast array of small-scale features all generated by the solution of the bedrock, formed predominantly by limestones

Karst aquifer — the assembly of ground water accumulated inside a karstic rock, enough to supply wells and springs

Karsted — rich in karst features, especially caves

Kyr — abbreviation for thousand years

Master joint — a tectonic discontinuity (fault line) that a given cave passage follows

Myr — abbreviation for million years

Neanderthal — believed by some to be an early human that was short, stocky, and stooped, with sloping foreheads, heavy eyebrows, jutting facces and bent knees

Orthokarst — karst formed on carbonate rocks mainly by solution

Parakarst — karst-like features formed on non-carbonate rocks, mainly by solution

Phreatic caves — flooded (water saturated) caves formed and/or located below the water table

Pseudokarst — karst-like features formed on any kind of rock by other ways than solution

Relict caves — caves without a flowing stream, which may have ponds or dripping water

Resurgences — the place where a sinking stream re-emerges to the surface

Scallops — spoon-shaped hollows in a cave wall, floor, or ceiling dissolved by eddies in flowing water

Sinkholes — funnel-shaped hollows, from a few feet to hundreds of feet in diameter

Speleothems — mineral deposits that form inside caves; especially stalagmites and stalactites

Troglobites — creatures which live only in caves (from Greek for "cave dwellers")

Troglophiles — creatures which spend some part of their life in caves (from Greek for "who like caves")

Trogloxenes — creatures that got into a cave by accident and which try to leave (from Greek for "foreign to caves")

Unidirectional air circulation — air flowing one way

Vadose caves — caves that formed and continue to exist mostly above the water table, the majority of their passages have air above the water

Xibalba — the Mayan "Underworld"

The Fossil Book - Glossary

4 C's — an aid for remembering four major events in Biblical history important to understanding fossils: God's perfect Creation, ruined by man's sin (Corruption), destroyed by Noah's flood (Catastrophe), restored to new life in Christ

Archaeology — the science that deals with human artifacts, and with things deliberately buried by humans

Arthropod — all creatures with jointed legs and a tough outside skeleton (exoskeleton) made of chitin: insects, crabs and shrimp, spiders, centipedes, and millipedes

Artifacts — products crafted by humans

Cambrian explosion — the sudden appearance of a wide variety of complex life forms in the lowest rock layer with abundant fossils (Cambrian); considered a challenge to evolution, these may be the first organisms in a corrupted creation to be buried in Noah's flood

Cavitations — bubbles formed by surging waters

Cephalopod — means "head-footed," since their tentacles come out of their heads, the most complex of all the invertebrates are the squid and octopus in the mollusk class

Creationist — one who thinks that (1) fossils show complex and separate beginnings because each kind was created well designed to multiply after kind, but that (2) fossils also show death, disease, and decline in variety and size because struggle and death followed man's sin (until Christ returns) and broght on Noah's flood

Diatoms — microscopic, one-celled plants whose walls are decorated with glass ($SiO2$) in exquisite patterns; mined and sold as diatomaceous earth, which is used in filtering and abrasion

Echinoderms — means "spiny-skinned," members of the starfish/sea star group usually have bony plates and spines embedded

Evolutionary series — a sequence of fossils that suggests how one kind of creature might have changed into others

Evolutionist — one who believes fossils will show that (1) millions of years of time, chance, struggle, and death changed a few simple life forms into all the complex and varied forms we have today, and that (2) new structures gradually developing from low to high in the geologic column will be seen when "missing links" are eventually found

Fossil — remains or trace of a once-living thing preserved by natural processes, most often by rapid, deep burial in waterlaid sediments

Gastropods — means "stomach-footed," since they walk on their stomachs; mollusk class which snails belong

Geologic column — a columnar diagram identifying rocky layers (strata) that form a sequence from bottom to top to indicate their relation to the twelve geologic systems

Geologic system — a major rock layer whose fossils are used to name it for one of the twelve groups in the geologic column diagram

Geology — the scientific study of the earth, including the materials that it is made of, the physical and chemical processes that occur on its surface and in its interior, and the history of the planet and its life forms

Index fossil — fossils used to identify a geologic system because they lived either (a) at a certain time or (b) in a certain place in the pre-Flood world

Invertebrate — animals without backbones

Living fossils — creatures found alive today that evolutionists thought became extinct millions of years ago

Malacology — a branch of science is devoted to the study of mollusk shells

Metamorphosis — the process of transformation from an immature form to an adult form in two or more distinct stages

Mollusks — a large phylum of animals with thick, muscular bodies and a complex system of organs

Nautiloids — fossils with tapered, chambered shells; some are coiled like the modern nautilus, others are curved like bananas, and still others are straight, like ice cream cones

Paleontology — the study of fossils

Palynology — the branch of paleontology that studies microscopic spores and pollen of plants

Paraconformities — a gap without erosion in the geologic column diagram; breaks the time sequence assumed by evolution, and may suggest fossils from different environments were rapidly buried by a lot of water, not a lot of time

Permineralized fossils — fossils preserved by minerals hardening in the pore spaces of a specimen such as a shell, bone, or wood

Petrified — fossils preserved by minerals completely replacing but preserving the pattern in the original wood, bone, etc.

Polystrates — fossils that cut through many layers, suggesting the sequence was laid down very rapidly

Protozoan — one-celled animal

Pseudofossils — false fossils; things that look like fossils but really aren't

Sediments — particles of sand, silt, clay, ash, etc. eroded and deposited by wind and water currents

Spicules — sponges that have hard skeletal structures of crystal-like spines

Splint bones — modern one-toed horses actually keep parts of the two flanking toes as important leg support structures (not useless evolutionary leftovers)

Stratigraphic series — sequence of fossils from lower to higher in the geologic column diagram (see above); thought to represent either (a) stages in evolution, or (b) stages in burial during Noah's flood

Stromatolites — banded rock deposits formed by blue-greens growing in mossy mats on rocks in the tidal zone along the shore; the mats trap and then cement sand grains to form a mineral layer, continually building new layers on top of earlier ones

Trace fossils — are not remains of plant or animal parts, but show evidence of once-living things

Trilobite — a crab-like creature that was the first fossil found buried in abundance around the world

Vertebrates — animals with backbones

Parent Lesson Plan

Now turn your favorite Master Books into curriculum! Each complete three-hole punched Parent Lesson Plan (PLP) includes:

- An easy-to-follow, one-year educational calendar
- Helpful worksheets, quizzes, tests, and answer keys
- Additional teaching helps and insights
- Complete with all you need to quickly and easily begin your education program today!

ELEMENTARY ZOOLOGY

1 year
4th – 6th

Package Includes: *World of Animals, Dinosaur Activity Book, The Complete Aquarium Adventure, The Complete Zoo Adventure, Parent Lesson Plan*

5 Book Package
978-0-89051-747-5 $84.99

SCIENCE STARTERS: ELEMENTARY PHYSICAL & EARTH SCIENCE

1 year
3rd – 8th grade

6 Book Package Includes: *Forces & Motion –Student, Student Journal, and Teacher; The Earth – Student, Teacher & Student Journal; Parent Lesson Plan*

6 Book Package
978-0-89051-748-2 $51.99

SCIENCE STARTERS: ELEMENTARY CHEMISTRY & PHYSICS

1 year
3rd – 8th grade

Package Includes: *Matter – Student, Student Journal, and Teacher; Energy – Student, Teacher, & Student Journal; Parent Lesson Plan*

7 Book Package
978-0-89051-749-9 $54.99

INTRO TO METEOROLOGY & ASTRONOMY

1 year
7th – 9th grade
½ Credit

Package Includes: *The Weather Book; The Astronomy Book; Parent Lesson Plan*

3 Book Package
978-0-89051-753-6 $44.99

INTRO TO OCEANOGRAPHY & ECOLOGY

1 year
7th – 9th grade
½ Credit

Package Includes: *The Ocean Book; The Ecology Book; Parent Lesson Plan*

3 Book Package
978-0-89051-754-3 $45.99

INTRO TO SPELEOLOGY & PALEONTOLOGY

1 year
7th – 9th grade
½ Credit

Package Includes: *The Cave Book; The Fossil Book; Parent Lesson Plan*

3 Book Package
978-0-89051-752-9 $44.99

CONCEPTS OF MEDICINE & BIOLOGY

1 year
7th – 9th grade
½ Credit

Package Includes: *Exploring the History of Medicine; Exploring the World of Biology; Parent Lesson Plan*

3 Book Package
978-0-89051-756-7 $40.99

CONCEPTS OF MATHEMATICS & PHYSICS

1 year
7th – 9th grade
½ Credit

Package Includes: *Exploring the World of Mathematics; Exploring the World of Physics; Parent Lesson Plan*

3 Book Package
978-0-89051-757-4 $40.99

CONCEPTS OF EARTH SCIENCE & CHEMISTRY

1 year
7th – 9th grade
½ Credit

Package Includes: *Exploring Planet Earth; Exploring the World of Chemistry; Parent Lesson Plan*

3 Book Package
978-0-89051-755-0 $40.99

THE SCIENCE OF LIFE: BIOLOGY

1 year
8th – 9th grade
½ Credit

Package Includes: *Building Blocks in Science; Building Blocks in Life Science; Parent Lesson Plan*

3 Book Package
978-0-89051-758-1 $44.99

BASIC PRE-MED

1 year
8th – 9th grade
½ Credit

Package Includes: *The Genesis of Germs; The Building Blocks in Life Science; Parent Lesson Plan*

3 Book Package
978-0-89051-759-8 $43.99

INTRO TO ASTRONOMY

1 year
7th – 9th grade
½ Credit

Package Includes: *The Stargazer's Guide to the Night Sky*; *Parent Lesson Plan*

2 Book Package
978-0-89051-760-4   **$47.99**

PALEONTOLOGY: LIVING FOSSILS

1 year
10th – 12th grade
½ Credit

Package Includes: *Living Fossils, Living Fossils Teacher Guide, Living Fossils DVD*; *Parent Lesson Plan*

3 Book, 1 DVD Package
978-0-89051-763-5   **$66.99**

INTRO TO ARCHAEOLOGY & GEOLOGY

1 year
7th – 9th
½ Credit

Package Includes: *The Archaeology Book*; *The Geology Book*; *Parent Lesson Plan*

3 Book Package
978-0-89051-751-2   **$45.99**

LIFE SCIENCE ORIGINS & SCIENTIFIC THEORY

1 year
10th – 12th grade
1 Credit

Package Includes: *Evolution: the Grand Experiment, Teacher Guide, DVD*; *Living Fossils, Teacher Guide, DVD*; *ParentLesson Plan*

5 Book, 2 DVD Package
978-0-89051-761-1   **$139.99**

SURVEY OF SCIENCE HISTORY & CONCEPTS

1 year
10th – 12th grade
1 Credit

Package Includes: *The World of Mathematics; The World of Physics; The World of Biology; The World of Chemistry; Parent Lesson Plan*

5 Book Package
978-0-89051-764-2   **$72.99**

NATURAL SCIENCE THE STORY OF ORIGINS

1 year
10th – 12th grade
½ Credit

Package Includes: *Evolutions the Grand Experiment; Evolution the Grand Experiment Teacher's Guide, Evolution the Grand Experiment DVD; Parent Lesson Plan*

3 Book, 1 DVD Package
978-0-89051-762-8   **$66.99**

SURVEY OF SCIENCE SPECIALTIES

1 year
10th – 12th grade
1 Credit

Package Includes: *The Cave Book; The Fossil Book; The Geology Book; The Archaeology Book; Parent Lesson Plan*

5 Book Package
978-0-89051-765-9   **$81.99**

ADVANCED PRE-MED STUDIES

1 year
10th – 12th grade
1 Credit

Package Includes: *Building Blocks in Life Science; The Genesis of Germs; Body by Design; Exploring the History of Medicine; Parent Lesson Plan*

5 Book Package
978-0-89051-767-3   **$76.99**

SURVEY OF ASTRONOMY

1 year
10th – 12th grade
1 Credit

Package Includes: *The Stargazers Guide to the Night Sky; Our Created Moon; Taking Back Astronomy; Our Created Moon DVD; Created Cosmos DVD; Parent Lesson Plan*

4 Book, 2 DVD Package
978-0-89051-766-6   **$110.99**

BIBLICAL ARCHAEOLOGY

1 year
10th – 12th grade
1 Credit

Package Includes: *Unwrapping the Pharaohs; Unveiling the Kings of Israel; The Archaeology Book; Parent Lesson Plan.*

4 Book Package
978-0-89051-768-0   **$99.99**

GEOLOGY & BIBLICAL HISTORY

1 year
8th – 9th
1 Credit

Package Includes: *Explore the Grand Canyon; Explore Yellowstone; Explore Yosemite & Zion National Parks; Your Guide to the Grand Canyon; Your Guide to Yellowstone; Your Guide to Zion & Bryce Canyon National Parks; Parent Lesson Plan.*

4 Book, 3 DVD Package
978-0-89051-750-5   **$112.99**

CHRISTIAN HERITAGE

1 year
10th – 12th grade
1 Credit

Package Includes: *For You They Signed; Lesson Parent Plan*

2 Book Package
978-0-89051-769-7   **$50.99**